ADAPTING AGRICULTURAL EXTENSION TO PEACEBUILDING

Report of a Workshop by the
National Academy of Engineering and United States Institute of Peace
Roundtable on Technology, Science, and Peacebuilding

Andrew Robertson and Steve Olson, *Rapporteurs*

NATIONAL ACADEMY OF ENGINEERING
OF THE NATIONAL ACADEMIES

UNITED STATES INSTITUTE OF PEACE

THE NATIONAL ACADEMIES PRESS
Washington, D.C.
www.nap.edu

THE NATIONAL ACADEMIES PRESS   500 Fifth Street, NW   Washington, DC 20001

NOTICE: This publication has been reviewed according to procedures approved by the National Academy of Engineering report review process. Publication of signed work signifies that it is judged a competent and useful contribution worthy of public consideration, but it does not imply endorsement of conclusions or recommendations by the National Academy of Engineering. The interpretations and conclusions in such publications are those of the authors and do not purport to present the views of the council, officers, or staff of the National Academy of Engineering.

The Roundtable on Technology, Science, and Peacebuilding, the sponsor of the workshop on which this report is based, is supported by funding from the U.S. Department of Defense (JDDM-3663-1), Qualcomm, National Science Foundation (ENG-1136841), U.S. Departmnt of Agriculture (59-0790-2-058), U.S. Department of State, and CRDF Global. Any opinions, findings, or conclusions expressed in this publication are those of the workshop participants.

International Standard Book Number 13:   978-0-309-25967-5
International Standard Book Number 10:   0-309-25967-3

Copies of this report are available from the National Academies Press, 500 Fifth Street NW, Keck 360, Washington, DC 20001; (888) 624-8373; online at www.nap.edu.

For more information about the National Academy of Engineering, visit the NAE home page at www.nae.edu.

Copyright 2012 by the National Academy of Sciences. All rights reserved.

Printed in the United States of America

THE NATIONAL ACADEMIES
Advisers to the Nation on Science, Engineering, and Medicine

The **National Academy of Sciences** is a private, nonprofit, self-perpetuating society of distinguished scholars engaged in scientific and engineering research, dedicated to the furtherance of science and technology and to their use for the general welfare. Upon the authority of the charter granted to it by the Congress in 1863, the Academy has a mandate that requires it to advise the federal government on scientific and technical matters. Dr. Ralph J. Cicerone is president of the National Academy of Sciences.

The **National Academy of Engineering** was established in 1964, under the charter of the National Academy of Sciences, as a parallel organization of outstanding engineers. It is autonomous in its administration and in the selection of its members, sharing with the National Academy of Sciences the responsibility for advising the federal government. The National Academy of Engineering also sponsors engineering programs aimed at meeting national needs, encourages education and research, and recognizes the superior achievements of engineers. Dr. Charles M. Vest is president of the National Academy of Engineering.

The **Institute of Medicine** was established in 1970 by the National Academy of Sciences to secure the services of eminent members of appropriate professions in the examination of policy matters pertaining to the health of the public. The Institute acts under the responsibility given to the National Academy of Sciences by its congressional charter to be an adviser to the federal government and, upon its own initiative, to identify issues of medical care, research, and education. Dr. Harvey V. Fineberg is president of the Institute of Medicine.

The **National Research Council** was organized by the National Academy of Sciences in 1916 to associate the broad community of science and technology with the Academy's purposes of furthering knowledge and advising the federal government. Functioning in accordance with general policies determined by the Academy, the Council has become the principal operating agency of both the National Academy of Sciences and the National Academy of Engineering in providing services to the government, the public, and the scientific and engineering communities. The Council is administered jointly by both Academies and the Institute of Medicine. Dr. Ralph J. Cicerone and Dr. Charles M. Vest are chair and vice chair, respectively, of the National Research Council.

www.national-academies.org

UNITED STATES INSTITUTE OF PEACE
Center of Innovation for Science, Technology, & Peacebuilding

The United States Institute of Peace is the global conflict management center for the United States. Created by Congress in 1984 to be independent and nonpartisan, the Institute works to prevent, mitigate, and resolve international conflict through nonviolent means. USIP operates in the world's most challenging conflict zones, and it leads in professional conflict management and peacebuilding by applying innovative tools, convening experts and stakeholders, supporting policymakers, and providing public education. The Institute translates its on-the-ground experience into knowledge, skills, and resources for policymakers, the US military, government and civilian leaders, nongovernmental organizations, practitioners, and citizens both here and abroad.

The Institute's permanent headquarters and conference center are located at the northwest corner of the National Mall in Washington, DC. The facility also houses the Academy for International Conflict Management and Peacebuilding and the Global Peacebuilding Center.

www.usip.org

WORKSHOP STEERING COMMITTEE

Pamela Aall (*co-chair*), Senior Vice President, U.S. Institute of Peace
Ann Bartuska (*co-chair*), Deputy Under Secretary, U.S. Department of Agriculture
Cathy Campbell, President and CEO, CRDF Global
Mark Epstein, Senior Vice President of Development, Qualcomm
Brian Greenberg, Director of Sustainable Development, InterAction
Mike McGirr, National Program Leader, NIFA-USDA, U.S. Department of Agriculture
Donald Nkrumah, Associate Director, Global R&D, Pfizer Animal Health
Riikka Rajalahti, Senior Agricultural Specialist, The World Bank

Staff

Geneve Bergeron, Research Assistant, U.S. Institute of Peace
Sheldon Himelfarb, Director, Center of Innovation for Science, Technology, and Peacebuilding
Greg Pearson, Senior Program Officer, National Academy of Engineering
Proctor P. Reid, Director, NAE Program Office
Andrew Robertson, Senior Program Officer, U.S. Institute of Peace
Ibrahim Shaqir, Interagency Professional in Residence, U.S. Institute of Peace
Frederick S. Tipson, Special Advisor, Center of Innovation for Science, Technology, and Peacebuilding

Acknowledgments

This summary has been reviewed in draft form by individuals chosen for their diverse perspectives and technical expertise, in accordance with procedures approved by The National Academies. The purpose of the independent review is to provide candid and critical comments to assist the National Academy of Engineering (NAE) in making its published report as sound as possible and to ensure that the report meets institutional standards for objectivity, evidence, and responsiveness to the study charge. The review comments and draft manuscript remain confidential to protect the integrity of the deliberative process. We thank the following individuals for their review of this report:

Gary Alex, Farmer-to-Farmer Program Manager, U.S. Agency for International Development
Mark Bell, Director, International Learning Center, UC Davis
Dennis Kopp, Retired Program and Analysis Officer, NIFA-USDA
Cindi Warren Mentz, Director, External Relations, Middle East/North Africa, CRDF Global
Ronald D. Oetting, Professor of Entomology, University of Georgia
Riikka Rajalahti, Senior Agricultural Specialist, The World Bank

Although the reviewers listed above provided many constructive comments and suggestions, they were not asked to endorse the views expressed in the report, nor did they see the final draft of the report before its release. The review of this report was overseen by Venkatesh (Venky) Narayanamurti, Benjamin Peirce Professor of Technology and Public Policy, Harvard School of Engineering and Applied Science, and director, Science, Technology and Public Policy Program, Harvard Kennedy School. Appointed by NAE, he was responsible for making certain that an independent examination of this report was carried out in accordance with institutional procedures and that all review comments were carefully considered. Responsibility for the final content of this report rests entirely with the authors and NAE.

In addition, both the National Academies and the U.S. Institute of Peace acknowledge the support of Under Secretary Catherine Woteki and Deputy Under Secretary Ann Bartuska in advancing this work. In addition to strongly advocating this study as a Roundtable activity, they supported the workshop and its planning by seconding Ibrahim Shaqir, Director of the Agricultural Research Program's Office of International Research Programs, to the Institute for four months. Mr. Shaqir's involvement was instrumental in shaping this project and its outcomes.

Contents

1 INTRODUCTION   1
   The Workshop, 3
   Discussion Themes, 4

2 CONFLICT IN RURAL SETTINGS   9
   Conflict over Land in Rural Settings, 10
   Postconflict Reintegration: Problems and Opportunities, 13
   Conflict Resolution Among Afghan Herders, 15
   Discussion, 17

3 EXTENSION SERVICES IN FRAGILE SOCIETIES   19
   Challenges, Needs, and Opportunities, 19
   Agricultural Extension in South Sudan, 22
   Agricultural Extension in Iraq, 25
   Discussion, 26

4 CAPACITY BUILDING AND TRAINING   29
   Skills, 29
   Legitimacy, 30
   Processes, 33

5 ORGANIZATIONAL CHANGE AND INSTITUTION BUILDING 35
   Options for Working with Ministries of Agriculture, 35
   The Need for Decentralization, 36
   Ensuring Sustainability, 38

6 TECHNOLOGICAL INFRASTRUCTURE 39
   Technological Capabilities for Extension and Peacebuilding, 39
   Potential Extensions of ICT, 40
   Involvement of the Private Sector, 41

7 FINAL OBSERVATIONS 43

APPENDIXES

A Agenda 45
B Attendees 49

1

Introduction

Societies have sought to improve the outputs of their agricultural producers for thousands of years.[1] In the 19th and early 20th centuries, efforts to convey agricultural knowledge to farmers became known as *extension services*, a term adopted from programs at Oxford and Cambridge designed to extend the knowledge generated at universities to surrounding communities.[2] Traditionally, extension services have emphasized a top-down model of technology transfer that encourages and teaches producers to use crop and livestock varieties and agricultural practices that will increase food production. More recently, extension services have moved toward a facilitation model, in which extension agents work with producers to identify their needs and the best sources of expertise to help meet those needs.[3]

Extension services can have a profound effect on the practices of agricultural producers and the agricultural productivity of nations. Many of these

---

[1] Jones, Gwyn E., and Chris Garforth. 1997. "The History, Development, and Future of Agricultural Extension." Chapter 1 in *Improving Agricultural Extension: A Reference Manual*, Burton E. Swanson, Robert P. Bentz, and Andrew J. Sofranko, eds. Rome: Food and Agriculture Organization of the United Nations.

[2] In most European countries, extension services are known as advisory services, and the two terms are used to varying degrees in other countries. This publication uses the term "extension services" to refer to all such activities.

[3] Swanson, Burton E., and Riikka Rajalahti. 2010. *Strengthening Agricultural Extension and Advisory Services*. Washington, D.C.: The World Bank.

services are widely disseminated and closely integrated into local communities, giving them a scope and influence not matched by more centralized programs. These features suggest that extension activities can contribute to peacebuilding in countries beset by conflict, albeit with organizational modifications and enhanced capacity in order for agents to engage in such activities effectively. Through the provision of agricultural and potentially peacebuilding information, extension agents can also strengthen the reputation and credibility of the government.

On May 1, 2012, the Roundtable on Science, Technology, and Peacebuilding held a workshop in Washington, DC, to explore whether and how extension activities could serve peacebuilding purposes. The Roundtable is a partnership between the National Academy of Engineering (NAE) and the US Institute of Peace (USIP). It consists of senior executives and experts from leading governmental organizations, universities, corporations, and nongovernmental organizations, was established in 2011 to make a measurable and positive impact on conflict management, peacebuilding, and security capabilities. Its principal goals are:

1. to accelerate the application of science and technology to the process of peacebuilding and stabilization;
2. to promote systematic, high-level communication between peacebuilding and technical organizations on the problems faced and the technical capabilities required for successful peacebuilding; and
3. to collaborate in applying new science and technology to the most pressing challenges for local and international peacebuilders working in conflict zones.

At a December 2011 meeting, the Roundtable agreed on a portfolio of high-impact peacebuilding activities in the following areas:

1. adapting agricultural extension to peacebuilding;
2. using data sharing to improve coordination in peacebuilding;
3. sensing emerging conflicts; and
4. harnessing systems methods for delivery of peacebuilding services.

Subcommittees are developing action plans for these areas; the May 1, 2012, workshop was the first in a series that will address the four topics. The Roundtable is committed to using these workshop activities as a basis for peacebuilding action in the field. Consequently, the long-term goal of each study area is to demonstrate a viable technical solution in a successful field trial.

Ann Bartuska, Deputy Under Secretary at the US Department of Agriculture and a co-chair of the Roundtable, explained during her introductory remarks that agricultural extension was chosen as the subject of the first meeting because of its focus on community-level change, which is a particular point of emphasis for the Roundtable.

The workshop and this summary are intended to (1) help policymakers think through the issues associated with the use of extension systems to stabilize rural societies after periods of war and (2) help managers of extension projects in postconflict environments design activities that promote peace.

THE WORKSHOP

Organization

Pamela Aall, Senior Vice President at USIP, Provost of USIP's Academy for International Conflict Management and Peacebuilding, and co-chair of the Roundtable, laid out the organization of the workshop. In the morning, two panel discussions featured speakers who explored the intersection of extension services and peacebuilding. The first panel looked at conflict in rural settings (Chapter 2), and the second examined the role of extension services in fragile societies (Chapter 3).

In the afternoon, the workshop participants divided into three groups to discuss specific aspects of extension services and peacebuilding. One group investigated changes in the skills of extension officers that could enable them to serve more effectively as peacebuilders (Chapter 4). The second looked at the corresponding changes required in the organization of extension services (Chapter 5). The third considered the technological infrastructure needed for extension officers to integrate peacebuilding into their activities (Chapter 6). The final session of the workshop featured reports from these groups and a summary of the workshop deliberations (Chapter 7).

Goal of the Workshop

The formal goal of the workshop was "to identify what peacebuilding activities could be delivered as components of existing extension services and what organizational modifications and new capabilities would be required to do so effectively." Or, as Aall put it, to answer two questions: Should extension services be used for peacebuilding purposes? If so, how should this be done?

Sheldon Himelfarb, Director of USIP's Center of Innovation: Science, Technology, and Peacebuilding, pointed out that peacebuilding activities can occur on a continuum of involvement and activism. At one end of the spectrum is a "do no harm" principle: activities must not exacerbate a conflict. Thus extension personnel are sensitive to the nature of the conflict and strive, through fairness and evenhandedness, not to make the conflict worse. At the other end of the spectrum is direct involvement in the driving forces behind a conflict—extension personnel may be active mediators in a conflict and work with opposing groups to reduce tensions. Between these two extremes, a wide range of activities may lend themselves to peacebuilding in a variety of contexts.

DISCUSSION THEMES

Several broad themes emerged during the workshop discussions. They are presented here not as consensus conclusions of the participants but rather as indicators of major issues that need to be examined when considering the possible roles of extension agents in peacebuilding.

How Can and Should Extension Personnel Contribute to Peacebuilding?

The broad role of extension agents, who act more as facilitators than as problem solvers, is to help agricultural producers gain access to knowledge, resources, and services that will increase their productivity and well-being. They can help build both social and agricultural capital in postconflict settings, and can help government agencies or nongovernmental organizations (NGOs) identify community needs for either development or security.

Extension agents may help manage conflict in rural communities in many ways. They can act as honest brokers between groups, providing guidance and information to assist in resolution of the conflict. They can organize producer associations or advise managers of shared resources to be inclusive and transparent in order to avoid exacerbating conflicts. By reducing conflict-related disruptions, they can enhance agricultural productivity and thus alleviate the material need that can drive conflict. Finally, their active presence in rural communities may enhance government credibility and encourage hope for a better future.

The roles of extension agents in both agriculture and peacebuilding vary greatly depending on the circumstances. In peacebuilding, the local causes of conflict define the issues an extension agent may confront in the same way that local agricultural issues determine the most useful forms of extension services. Conflict issues in which agents may have a role include land disputes, disputes between herders and pastoralists, and reintegration of former combatants and displaced people in communities. Training in conflict analysis was identified as a necessity for peacebuilding work.

Extension agents already have a full slate of responsibilities, and adding peacebuilding activities could easily be overwhelming. A role in peacebuilding therefore needs to be integrative and not additive. However, agents should already be engaging in activities that both directly and indirectly can serve peacebuilding purposes. They should act as brokers of information and access to information among groups and between groups and the government. (Unfortunately, however, extension agents often lack the skills and resources to function as brokers.) They provide services that both increase agricultural productivity and enhance the economic security of agricultural producers and can serve as peacebuilders through these and other extension activities.

In postconflict environments, extension agents must be highly conscious of the possibility of their exacerbating tensions in the communities they serve by directing extension services and support in ways that exclude groups on the basis of race, ethnic identity, class, gender, or education.

Finally, in rural communities, much agricultural work is done by women. Therefore, extension systems designed to support both agriculture and peacebuilding would show greater promise if programming specifically engaged rural women.

How Should Extension Agents Be Selected, Trained, and Motivated?

Extension agents need a very wide range of skills to do their jobs well, from technical knowledge to a variety of social skills. Peacebuilding adds to this list an ability to analyze conflicts. Extension personnel need to understand the drivers of conflict and the likely consequences of their actions.

To be effective, extension agents need to be respected, trusted, and accepted by their clientele, regardless of their level of education or group affiliation. To that end, their advice needs to be objective, useful, and nonpartisan. Furthermore, they need credible sources of information, continuously updated skills, and trustworthy partners. Extension personnel also gain legitimacy by working with people who are trusted in the community. If extension personnel are motivated only by a paycheck or having a government job, their legitimacy will be questioned.

In some cases, extension agents may be more likely to gain trust if they are from a local area and are provided with training. However, the effectiveness of such agents may be compromised if they are part of an elite or associated with a particular side or agenda in a conflict.

What Institutional Changes Are Needed to Support a Peacebuilding Role for Extension Agents?

Extension services typically operate in ministries of agriculture, and changes in ministry organization may support peacebuilding as part of extension activities. For example, ministries and extension services may become explicitly involved in conflict analysis, especially when conflicts affect or are affected by agriculture. Or a ministry and its extension officers may become involved in the reintegration of former combatants or regions of countries into the broader society.

In many countries, the extension capacity of ministries of agriculture has been severely limited by long-term underinvestment in staffing, training, and programming. This problem has been compounded by an approach to extension that tends to be centralized and top-down. In these cases, much greater decentralization, with a capacity to support local grassroots extension activities, has occurred through the activities of NGOs. In peacebuilding efforts to address communities' expressed needs, such bottom-up approaches may have the desirable effect of improving both agricultural productivity and social stability.

Although extension systems have become pluralistic in nature, with services provided not only by government but also NGOs and private organizations, from a conflict perspective it is important for extension activities to build government technical capacity and political credibility. Similarly, although sustainability is often likelier when support comes from multiple sources—public, private, governmental, or nongovernmental—government support is specifically necessary. And importantly, to be effective in their work, extension agents need to have the necessary support, resources, and tools, including appropriate salaries, incentives, operating budgets, training, and evaluation programs.

In the United States, academic institutions involved with agricultural extension have a curriculum tightly linked to agricultural research, but this is often not the case in developing countries generally and postconflict societies in particular. Universities can be an excellent source of training for extension personnel, but weak links with universities in developing countries detract from training and access to science-based information generated through sound and appropriate research.

How Can Technology Support a Role for Extension Agents in Peacebuilding?

Information and computer technology (ICT), the technological area most likely to have an immediate impact on peacebuilding, is rapidly becoming cheaper and more powerful. Advances in ICT have significant and growing potential to improve agent access to information and expertise for use in both agricultural extension and peacebuilding.

The technologies need to be trusted and the information provided valid. More specifically, ICT should be inexpensive and easy to use, support long-term capacity to improve both agricultural productivity and social stability, and broaden access to information for all groups in agricultural communities. In addition, the technologies need to be neutral in their application and usable among groups without much formal education.

It is particularly helpful to encourage and support communities in determining how best to use technology to solve problems and meet needs, including in ways perhaps not originally envisioned. For example, in addition to conveying information between farmers and extension agents, cell phones can register images and are therefore useful when documented evidence is required.

Enabling an extension agent to provide information in response to a farmer's question quickly builds trust in the individual agent and enhances the credibility of the larger extension system. Cell phones are often the best way to deliver information easily and inexpensively.

2

Conflict in Rural Settings

Conflict affects agricultural communities in multiple ways. Disagreements over rights to land, water access, and water quality can act as flashpoints, and in the aftermath of conflict those who return, whether refugees or demobilized soldiers, may create further conflict by increasing demand and thus stress on a community's economic and social capacity.

Extension agents can help to prevent or reduce conflict, as described by three speakers in the first session of the workshop. The presenters considered the potential roles of extension agents in conflicts over land in rural settings, challenges associated with postconflict reintegration, and experiences providing training for mediating disputes between farming and pastoral communities in rural Afghanistan.

Several possibilities emerged from the speakers' remarks. Speakers observed that extension agents can act as honest brokers between groups or between a group and the government. Agents can provide information—or access to information or other resources—that, directly or indirectly, reduces conflict. They can provide a variety of services, such as training or organizing producer associations, that can serve both agricultural and peacebuilding purposes. Through these and other means, extension personnel can promote peacebuilding, with the understanding that transparency and accountability are essential in all activities to avoid the appearance of favoritism and to foster trust.

CONFLICT OVER LAND IN RURAL SETTINGS

Wars often involve land, said Jon Unruh, Associate Professor of Human Geography and International Development at McGill University. In fact, according to the United Nations War-torn Societies Project, in 40 percent of postconflict countries clashes eventually resume, and land is the leading cause.

There are numerous reasons for land-related conflicts. Groups may struggle for control of lands with high-value resources, such as diamonds, timber, minerals, or cash crops. The identity of individuals, tribes, or factions may be attached to land. Wars may involve forced dislocation, land confiscations, or legalized evictions. Deeply held grievances that are not resolved by a peace accord may be related to land issues. Displaced people may return to areas that are occupied by others, endangered by land mines, or agriculturally damaged. Returnees may have little ability to prove their claims to land, and opportunists may make claims with little justification.

Land tenure in crisis situations is very different than in stable settings, Unruh said, as are solutions. What may work well in stable, peaceful settings can be very difficult to implement and enforce in societies recovering from war. People may lack fair access to courts or knowledge of the law and their legal options. People may pursue their land rights in aggressive or confrontational ways. These and other factors can lead to a buildup of competition, inequity, grievance, resentment, animosity, and violence.

Informal and Formal Legal Systems

A major problem, said Unruh, is that countries beset by conflict often do not have legitimate and fair ways of managing disputes through their legal systems. After a war, the state may not be trusted because it took one side during the conflict. Institutions may have collapsed, including the judicial system. Deeds, titles, and records are vulnerable to destruction, disorganization, looting, and fraud.

In such cases, informal or customary land rights may conflict with other forms of land tenure. Without a way to be legally validated, the customary tenure may degrade, collapse, or be abusively manipulated in a crisis situation. It then becomes a major challenge to establish, reestablish, secure, defend, prove, or confront claims to property, land, or territory, often in parallel with the splintering of society into postwar communities bound by factors such as dislocation, identity, ethnicity, or religion.

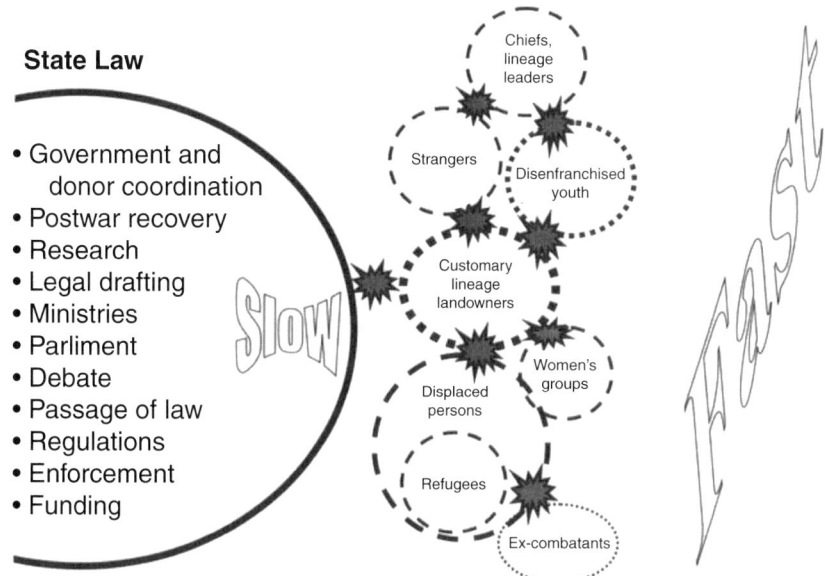

FIGURE 2-1 Postwar societies are typically divided between a formal system of state law that operates on a slow timescale and a fractured assemblage of groups that operate on fast timescales, with conflict (denoted by starbursts) between and among these systems and groups. SOURCE: Unruh workshop presentation.

A fundamental need in such situations is to connect informal legal pluralism in postwar scenarios with formal law. Informal legal pluralism operates quickly, does not wait for formal legislation, and functions in an isolated manner (Figure 2-1). Formal law, in contrast, operates slowly and depends on a complex set of institutions. It can also be confrontational, both internally and with the diverse actors common in fractured postwar societies.

The Need for an Honest Broker

What is needed in this situation, according to Unruh, is someone who can broker differences both within and between the formal and informal systems (Figure 2-2). This actor should be present in the rural area but not seen as an agent enforcing the power of the state. Although extension personnel may be agents of the state, they lack the authority to enforce—their

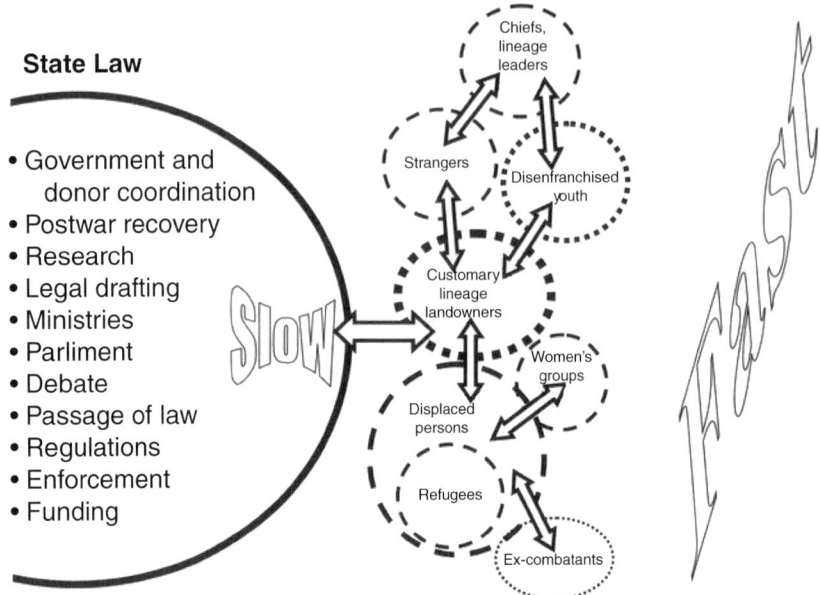

FIGURE 2-2 Agricultural extension agents can act as honest brokers between the formal and informal systems and among groups in a postwar society, as represented by the double-headed arrows. SOURCE: Unruh workshop presentation.

function is primarily educational—and so may be more easily accepted by local communities.

Agricultural extension agents may know where to go and whom to see in government to start the process of dealing with a complaint, dispute, or registration, said Unruh. They may know which government offices or staff to avoid or work around. They can serve as a go-between among factions or between a particular group and the government. Agents from the regions where they are working may have local contacts and be familiar with local needs and opportunities.

These brokers, whether extension agents or other individuals, need to be able to package evidence to be usable under state law. They need to be familiar with customary means of claiming land tenure and know how to upgrade them to more formal claims. In this respect, an important function is to encourage processes that make customary institutions relevant to state law. The broker may thus combine evidence such as agricultural improvements, oral accounts, inheritance claims, past allocations by traditional leaders, long-term use or occupation of land, and other means to make a more formal land tenure claim. As a particular example, Unruh cited the use of a mobile phone to take a picture of a grandfather's tomb or a boundary marker to serve as evidence. Honest brokers of information can also help to make parts of state law understandable and workable for local communities.

Such assistance in upgrading claims from informal or indirectly relevant documents, Unruh observed, encourages a transition from legal pluralism to more formal methods of making and appealing decisions.

Use of extension agents as facilitators of dialogue in land disputes could be faster and more effective than edicts from national capitals, Unruh said, although he added that it is important not to overburden extension agents with responsibilities.

A more purposeful and pervasive effort can take advantage of existing beneficial systems and build on them, bearing in mind that what is possible will likely differ in a risk-averse postconflict society versus one that is ready for development.

POSTCONFLICT REINTEGRATION: PROBLEMS AND OPPORTUNITIES

Caroline Hartzell, Professor of Political Science at Gettysburg College, discussed two aspects of the reintegration of individuals and groups into societies after conflicts. The first centers on reintegrating former combatants into civil society—the last phase of disarmament, demobilization, and reintegration (DDR).

Reintegration of Combatants

A major problem, said Hartzell, is that the reintegration phase of DDR tends to get much less emphasis and funding than the other two phases. But a failure to reintegrate former combatants can pose very serious problems, including the resumption of conflict. In Angola, for example, repeated peace settlements have failed because combatants stay armed and remain in the bush. Failure to integrate armed combatants also poses problems for societies when members of these groups prey on communities, engage in illegal economic activity, or maintain connections to former commanders. Even if such activities do not lead to a recurrence of conflict, they threaten the quality of peace.

Many communities are reluctant or unwilling to reintegrate former combatants back into the community, so DDR efforts have begun to address this. Reintegrated members of communities need to make a living but often have few skills and little or no capital. The agricultural sector is an obvious possibility for earning a livelihood, especially since many former combatants are from rural areas. DDR initiatives have sought to encourage this option through the provision of microcredits, training, or other forms of assistance. In Liberia, for example, ex-combatants in regions susceptible to the resumption of conflict received several months of training, psychosocial counseling, and a start-up agricultural package. They also had the opportunity to resettle in different areas, which helped break up combat-oriented networks.

A comparison of ex-combatants in Liberia who went through the program with a control group found that the former had a higher engagement in and commitment to agriculture. They also experienced a rise in durable wealth, although they had little actual change in income, which was not surprising, said Hartzell, given the ebbs and flows of the agricultural cycle. In addition, they experienced some improvement in social engagements, citizenship, and stability. However, their engagement in illicit or illegal economic activities did not change. This was a period of high gold prices, Hartzell noted, which led to considerable illegal mining. The treatment group continued to engage in this illegal activity but devoted fewer hours to it because of their commitment to farming.

Reintegration of Communities

Communities and parts of countries often need to be reintegrated into the state after a conflict. If a region was marginalized or ignored by the central government, its residents may feel little loyalty to the state and instead support nonstate actors challenging the government. They also may become involved in the illicit economy. Hartzell pointed to Afghanistan and Colombia for examples of such regions; in Colombia, entire regions of the country were controlled by nongovernmental entities and devoted to coca production, with an attendant loss of other agricultural skills and knowledge.

In these cases, it is necessary to reintroduce government authority in areas where it has been minimal or absent for an extended period, reintegrate the region into the nation, and replace illegal economic activities. In Colombia, for example, a program funded in part by the U.S. Agency for International Development's (USAID's) Office for Transition Initiatives provided technical assistance to agricultural producers to help them make the transition from the illicit economy to legal income-generating activities, such as the production of milk and pork for sale in local markets. The program emphasized community participation in defining needs and priorities, and worked with existing producer associations and encouraged the formation of new ones. These actions fostered a sense of citizenship and empowerment among the formerly marginalized communities, said Hartzell, and agricultural extension services were a central component of the effort. In addition, community building led the occupants of these regions to think of themselves as Colombians once again and not to be denigrated for their production of coca.

Programs need to be sustainable, said Hartzell, which means the country's government needs to be committed to them, even if occupants of cities are not enthusiastic about government investments in rural areas. Such investments can increase land values and give rise to tensions if claims to land are not secure or if people are pressured to sell their land. Furthermore, metrics of success in these programs are not easy to devise or assess, which poses a challenge to the commitment necessary for sustainability. The most appropriate and effective model will differ depending on the context.

CONFLICT RESOLUTION AMONG AFGHAN HERDERS

The nomadic herders in Afghanistan known as the Kuchi demonstrate a fundamental lesson of the application of extension to peacebuilding, said Michael Jacobs, Co–Principal Investigator and Chief of Party for the Afghanistan Pastoral Engagement, Adaptation and Capacity Enhancement (PEACE) project. Initially, Jacobs, a range ecologist, came to Afghanistan to improve livestock production. But he and his colleagues quickly understood that improved land management or veterinary practice was not possible for the Kuchi without addressing the conflict issues that limited their ability to manage land use. Being able to resolve conflict and negotiate passage to new pasture was the principal barrier to improvement in Kuchi productivity.

The Conflict: Sources and Effective Approaches

A survey of Kuchi herders showed that insecurity along migration routes was the number one risk for their livelihood. Jacobs explained that if the herders cannot get their animals to the mountains to graze, they cannot make a living, the sheep and goats suffer, and so does the rangeland on which the herds graze.

The conflicts that have arisen along migration routes, driven partly by population expansion and land conversion, are very complicated, Jacobs noted. Some are politically motivated; others result simply from a lack of communication. Most important, relationships between herders and villagers were very poor after years of war. But leaders of both groups have been united in wanting peace, which has been encouraging.

The PEACE project approached the challenges along migration routes by training and supporting Kuchi leaders to resolve land conflicts for their people. Project staff also sought to reestablish the relationship between villagers and herders, in part by seeking out local village and Kuchi leaders who would work together to resolve conflicts and build peace. Under the Independent General Director of Kuchi, 31 provincial directors are responsible for representing and assisting Kuchi communities. As part of this effort, young Kuchi leaders have been intensively trained to initiate peace *shura*s (consultations) in their communities. In addition, the PEACE project has been working with the Sanayee Development Organization, a local NGO experienced in delivering culturally appropriate peacebuilding and conflict resolution skills that go beyond the traditional methods used in Afghanistan.

As word spread of the PEACE project's successes, President Hamid Karzai's adviser on tribal affairs asked the project's leaders to try to solve a particular issue related to Kuchi and Hazara in Wardak province. The leaders agreed, but in return asked to expand the project to other migration areas. Today, said Jacobs, 75 peace ambassadors, including both village and Kuchi leaders, are working in seven regions of the country to resolve land and resource conflicts, including more than 900 conflicts in the past year. To build sustainability into the program, young Kuchi leaders are being trained to teach other Kuchi the PEACE project mediation and communication techniques.

Lessons Learned

Jacobs drew several lessons from his experience with the PEACE project. First, it is better to facilitate a conflict resolution effort than to appear to be directing it. PEACE took this route by partnering with a local Afghan NGO to implement a training curriculum. Related to this, Jacobs emphasized the importance of working with people whom the community already trusts. For example, working with local people may be preferable because they are trusted more than are people from the central government. Often, he said, people working in Afghanistan do not pay sufficient attention to this.

Outside groups coming into an area need to pick their partners wisely, especially government partners, Jacobs said, as they can make the difference between a program that is sustainable and one that goes nowhere.

Finally, development projects that involve agriculture and natural resource management can create as well as resolve conflicts. These projects need to be thoughtfully designed and provide services equitably to avoid doing more damage than good.

DISCUSSION

Three broad topics emerged from the participants' comments: (1) the tasks expected of extension agents, (2) the need for trust between extension personnel and the people they serve, and (3) the most effective model for extension services.

In response to a question about the potential number of different roles for extension agents, Unruh spoke of extension agents as a "user-friendly doorway" to alternatives to violence, offering advice or information or providing contacts to people in government or other organizations. Agents cannot necessarily be general purpose problem solvers, he said; if, for example, they become judges in disputes, they can become connected to power brokers in ways that are problematic. But in a bottom-up extension approach, information brokering and facilitation roles can be part of an agent's job description, and peacebuilding activities can be integrated into this role rather than being taken on as additional responsibilities.

Siddhartha Raja, Analyst with The World Bank, observed that extension agents should focus either on rights and laws or on economic development, and that trying to do both may lessen their capabilities in each area. Jacobs reiterated that extension agents must be careful not to exacerbate a conflict—for example, by failing to deliver services equitably. Cindi Warren Mentz, Director, External Relations, Middle East and North Africa, for CRDF Global, returned to the fundamental observation that extension agents provide jobs and contribute to stability by building agricultural capacity and increasing productivity.

Closely related to the tasks of extension agents is the nature of their relationship with the local population. Jacobs observed that extension personnel must be trusted to be effective. Being from the community they serve can contribute to this trust. Unruh, too, pointed to the advantages of training people from local communities to be extension agents. These people may not have the agricultural expertise of an outside expert, but they will be well respected and legitimate in that setting. Mike McGirr, National Program Leader for the National Institute of Food and Agriculture at the U.S. Department of Agriculture (USDA), confirmed that to change behaviors, an extension agent needs to be seen as a credible person by the community, and even more so when dealing with contentious issues.

However, Gary Alex, Farmer-to-Farmer Program Manager for USAID, cautioned that extension agents from local areas may be from an elite group or have social links on one side of a conflict that make them less trusted. In that case, extension work done by an agent from a different region, an NGO, or some other neutral body may be more effective.

Unruh pointed out that a community just emerging from war typically has very different needs from one that is several years removed from a crisis. Immediately after conflict, a country or region may be very risk averse and focused on not making things worse, whereas economic development often requires that a population be willing to take some risks, whether trying a new variety of seeds or accepting an extension agent's advice. Thus an extension agent may be able to engage in more traditional activities in the latter situation and may need to devote more attention to peacebuilding and stability in the former.

Hartzell added that the capacities of extension personnel differ greatly from country to country. In Liberia, for example, few people are available to serve as extension agents, whereas more people have those skill sets in Colombia. Unruh added that a significant challenge can be to convince a person with a university degree to serve as an extension agent in a war-torn part of a country. In such situations, a more effective approach is to identify the local problems that need to be solved and work toward local solutions.

Hartzell concluded that the adoption of a single model in all cases is not appropriate. Rather, the nature of the conflict and the capacity that exists in a country or region will determine what kinds of extension services will be most effective. And the model adopted for those services will affect both the tasks in which extension personnel engage and the degree to which they are trusted.

3

Extension Services in Fragile Societies

Extension agents working in communities affected by conflict face challenges beyond those normally associated with their jobs. Conflict may have prevented them from acquiring the background, training, or motivation needed to do their job well. They may not have the resources needed to make agricultural improvements. The societal dividing lines created by conflict may limit the cooperative activities on which extension is based.

During the second session of the workshop, three speakers analyzed these challenges and ways of overcoming them. It was clear from these presentations that surmounting barriers to successful extension in fragile societies almost always requires conflict management, which opens multiple routes for peacebuilding tied to extension activities.

CHALLENGES, NEEDS, AND OPPORTUNITIES

Agricultural extension, whether in fragile or secure societies, can be defined as the provision of knowledge to agricultural producers so that they will make a positive change, said Mark Bell, Director of the International Learning Center at the University of California, Davis, College of Agriculture and Environmental Sciences. This knowledge needs to be credible, as does the person who delivers it. Channels need to be available for the transmission of knowledge both from agents to agricultural producers and from producers to agents. All these conditions must be met for producers to make positive changes in practice with the information they receive.

With these requisites in mind, Bell analyzed the potential for agricultural extension in fragile societies in terms of challenges, needs, and opportunities.

Challenges

Several challenges are common in developing countries. For example, farmers are innovative and smart, said Bell, but they are not necessarily literate. The literacy rate for males in Afghanistan is about 40 percent, so knowledge often must be conveyed through means other than writing.

In addition, the economics of farming in developing countries poses challenges. Many farmers do not have ready access to credit or agricultural inputs, and the size of their farms is often small. In developed countries, an extension agent can talk to one farmer and have an influence over large areas. In developing countries, the agent must reach many more farmers. In addition, the agricultural infrastructure and markets in developing countries may be less robust than in developed countries.

Finally, in developed countries such as the United States, the institutions involved with agriculture and agricultural extension are tightly linked (Figure 3-1). In particular, major components of research, education, and exten-

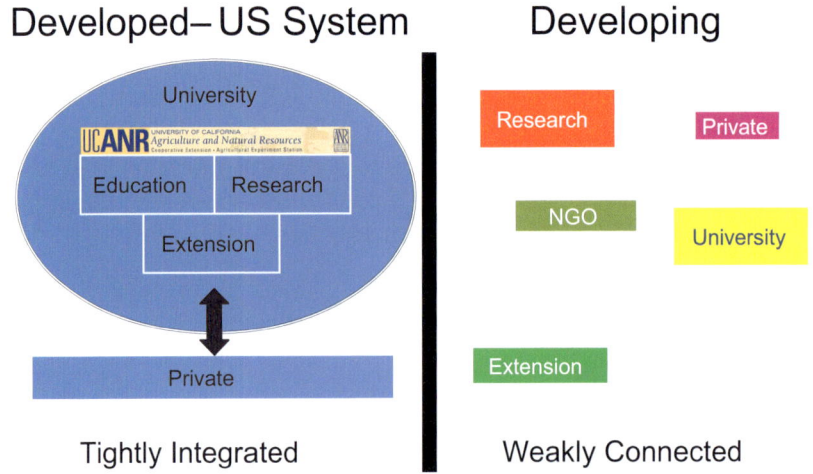

FIGURE 3-1 In developed countries such as the United States, extension systems are strongly tied to research and education in universities and to the private sector, whereas in developing countries these institutions tend to be largely separated. SOURCE: Bell workshop presentation.

sion are based in universities, which may have strong connections with the private sector. In contrast, linkages are much less strong in developing countries, meaning that institutions work largely in isolation from each other.

Needs

The number one need for successful extension, said Bell, is technical knowledge, which has to be credible and unbiased to win acceptance by agricultural producers. In their role as agricultural experts, extension agents provide farmers with objective, neutral advice based on science.

In addition, in their role as peacebuilders, extension agents must ensure that their activities do not exacerbate conflict. Bell suggested a number of desirable technical and personal skills for extension agents (Box 3-1). Agents also need to have the personal rapport to apply these technical skills in the field.

BOX 3-1
Desirable Skills for Extension Agents

- Team building
- Concept development
- Change management
- Delegation
- Conflict resolution
- Communication
- Planning
- Project management
- Facilitation/mediation
- Priority setting
- Time management

Successful extension activities require participatory approaches, Bell said. Producers have considerable local knowledge that needs to feed in to the extension process, both because of the way this knowledge interacts with the information an extension agent provides and because of the value of this knowledge to other producers.

Extension should take a process-driven approach, according to Bell, in which consideration of audiences and needs leads to solutions, the development of core messages, the delivery of those messages in an accessible form, and evaluation of outcomes. The process should start at the level of the producers rather than through top-down directives.

Finally, institutional elements such as salaries, training, evaluation, and motivating forces are necessary for success. Extension agents need to remain engaged and motivated, despite the institutional fragmentation that is characteristic of many developing countries.

Opportunities

If challenges can be overcome, agricultural extension has the potential to make an important difference in the lives of agricultural producers, their families, and the people who depend on those producers, Bell observed. But he reiterated that extension services will vary depending on local conditions. Extension also needs to draw on a diverse array of potential participants, and their availability will vary from place to place.

Extension services can contribute to peacebuilding, Bell concluded. The simultaneous challenge and opportunity is to bring together people interested in both extension and peacebuilding and build bridges between the two activities.

AGRICULTURAL EXTENSION IN SOUTH SUDAN

Jim Conley, Senior Agriculture Adviser in the Civilian Response Corps of the U.S. Department of Agriculture, has been an extension agent in the United States and has worked on reconstruction and stabilization projects in South Sudan, Iraq, and other countries. Most recently, he has been working in South Sudan's Jonglei State, which is about the size of Pennsylvania. The state has six different tribes, and about 60 indigenous languages are spoken throughout the country. Conflicts are common between farmers and pastoralists and among other competing groups. The major roads in the state are either dirt or gravel and often are impassable during the rainy season.

Conley has had an office in the Ministry of Agriculture and has observed the development of the extension service in Jonglei State. The goal of the state is to have 24 extension officers in each of the state's 11 counties, which is a large staff for the state. According to Conley, the Ministry of Agriculture has a larger geographic and personnel footprint than any other branch of government.

Extension officers are expected to speak the local language, which usually means that they are from the area. This contrasts with the practice in the United States, where new extension agents typically work in areas other than their local area so that they do not bring preconceptions or biases to their jobs. In Jonglei State the typical education level of an extension officer is primary or secondary school; very few have university degrees. New extension officers receive three months of intensive training through the NGO Norwegian People's Aid, with an additional three months of training for those who do well.

Obstacles to Success

Extension services in South Sudan face a number of serious constraints, said Conley. The country has been experiencing open acts of war, which have caused loss of life, property, productivity, opportunity, and social capital. The government's budget relies heavily on oil, but production was shut down due to conflicts over oil transport. Resources are almost nonexistent, with no money for even basic supplies or technologies. Extension officers include "deadwood" such as men who fought in the army for many years and found nonmilitary jobs with the government; even if these men had agricultural skills in the past, they are likely to have lost them, and most have little familiarity with computers or other technologies. Extension organizations in South Sudan have virtually no academic connection to universities, and any connections that do exist are informal. And extension officers are accountable to the Ministry of Agriculture or to their direct supervisors, Conley said, not to the people they serve, whereas the flow of accountability should run in the opposite direction.

Potential Roles for Extension in Conflict Mitigation

Conley described multiple ways for extension services to contribute to conflict mitigation. Agents can, for example, take steps to promote and reinforce community policing by bringing in experts with the right kind of technical knowledge to foster partnerships for community safety and other safety-enhancing initiatives. They also can spur community and economic development through agricultural and other improvements. And they can catalyze progress on environmental issues, again by bringing in experts who can provide information and engage in dialogue to resolve differences and arrive at solutions.

By way of illustration, Conley cited the Democratic Republic of Congo, where each district by law has a community agriculture council. The extension director is chair of the council, which may include a few other government officials, but most of the council members are farmers. These councils can guide extension services, much as advisory committees do in the United States. For example, when illegal checkpoints began to appear where fees were extorted from farmers to transport their goods, the community agriculture councils devised a plan for farmers to call someone who could relay the information about illegal checkpoints to law enforcement. Thus the farmers identified a need and the extension service figured out a way to meet that need.

Potential Roles for Extension in Peacebuilding

Agricultural extension in South Sudan could mitigate conflict by contributing to social capital through brokering and bridging functions or by providing early warning of emerging conflict through the assessment and monitoring of developing situations. Agents could provide early warnings about incipient conflicts, serve as honest brokers by providing information or enlisting the help of experts, and work directly with competing groups to resolve conflicts.

All of these options require training in both technical and social skills, and Conley cited several possible models for such training. First, the three months of agricultural training for extension officers could include training in conflict management and group facilitation, perhaps in partnership with an NGO that has experience and expertise in that area. Another useful option would be for the government to have a ministerial specialist in conflict resolution. Alternatively, when extension officers find themselves working on highly politicized topics, they could call in agents from other parts of the country to ensure that the extension system remains neutral.

In rural communities, much agricultural work is done by women, who Conley said have made some strides especially in improving their status working with local and international NGOS. In fact, he surmised that an extension system with peacebuilding as a component would hold more promise if it engaged rural women specifically.

Other local capacity should also be tapped. An example of how to cultivate and apply local skills and expertise, Conley said, is the Barefoot College, an NGO in India that uses local knowledge to make rural communities self-sustaining through development activities owned and managed by those in the community itself.

South Sudan has strengths, Conley observed: an energized youth who are eager to contribute to the country, people returning to the country who want the nation to succeed, and a growing emphasis on women's empowerment. With sufficient training and resources, extension services could draw on these and other strengths to play a substantial role in peacebuilding in South Sudan, he concluded.

AGRICULTURAL EXTENSION IN IRAQ

David Nisbet, Supervisory Microbiologist at the USDA Agricultural Research Service, described his experience as an agricultural adviser on a provincial reconstruction team in Iraq. He worked in Karbala Province on the western bank of the Euphrates River in a region of great conflict, where the convoys in which he traveled were often attacked. It was not a matter of Sunni–Shiite fighting but rather of intertribal competition for resources in the province.

At first local officials in the province would not interact with Nisbet, partly because of the failure of a promised transnational water pipeline project. Eventually he was able to make contact with the agricultural extension service in the province, and he found the people there to be well educated and sophisticated. The director of the office, in particular, was very effective in working across tribes and was committed to the community.

The province had a very active vegetable farming industry, and the extension office was working hard to deliver new technologies to farmers. Surprisingly, given the socially conservative society, the extension agents aggressively recruited women to the programs. In fact, Nisbet said, it was quite likely that the programs focusing on women were among the more effective that the agency funded. But Nisbet characterized the level of the technology as comparable to that of 1950s American agriculture, and resources were not available to make needed improvements.

Unintended Consequences of Good Intentions

Nisbet cautioned that efforts by outside groups to prepare extension personnel to undertake peacebuilding activities may not be appropriate or accepted in areas of conflict. In such situations, participating locals in positions of authority have to be excellent politicians, as was true of the agricultural extension officer with whom Nisbet interacted who did not become a victim of the violence gripping the country. Others were not so fortunate. Another extension agent with whom Nisbet worked disappeared for six weeks because he had been working closely with the United States, and women with whom the United States was working were beaten.

Good intentions can have other unintended consequences, Nisbet said. Iraqis did not necessarily see themselves as involved in a conflict until they became involved with the United States. He reported that the people with whom he worked would have left Iraq if they could, because essentially they became mice in a cat-and-mouse game.

In addition, large quantities of money were misspent. One of his major accomplishments, Nisbet said, was to block the development of a large poultry industry, which would immediately have failed in the 125° heat of Iraq's summers.

Potential Roles for Agricultural Extension

Agricultural extension in Karbala Province could have a huge role, said Nisbet, by making the province into an exporter of agricultural products. To be effective in improving either agriculture or peacebuilding efforts, extension officers need training, Nisbet stated—not necessarily in the United States, but perhaps in neighboring countries where they could learn what would work effectively in Iraq. Finally, the agricultural sector needs 21st century technology if it is to achieve its potential.

DISCUSSION

The discussion following these presentations revolved around the broad subjects of roles, metrics, and motivation.

Fred Tipson, Special Adviser at USIP, commented on advantages and considerations related to the many different roles of an extension agent. For example, someone from outside a community may not be connected to local disputes. In some places, an extension agent may serve as a community organizer, whereas in other places that role would be inappropriate or perhaps even dangerous. Different roles call for different skills—whether those of a diplomat, technologist, or anthropologist—and will also be influenced by the partnerships that often are required to create change. For example, a major challenge can be getting local agricultural producers to work with an extension agent, and different approaches may be more—or less—effective in different settings for achieving that end.

Nisbet reiterated the potential role of extension personnel as sentinels for local developments, whether related to agriculture or other activities. Agents can convey information to appropriate institutions for action without being seen as personally responsible for the action.

Jacqueline Wilson, Senior Program Officer at USIP, observed that extension personnel should be "connectors"—for example, connecting people with knowledge to people with the leverage to get things done. As a specific example, people in the community may be excellent agriculturalists—as Judith Payne, e-Business Adviser at USAID, observed—and extension agents should tap into their expertise. Agents also can convene parties with diverse interests in searching for common ground.

Montague Demment, Associate Vice President, International Development, for the Association of Public and Land-grant Universities, called attention to the difficulty of developing metrics to assess the value of investments in both agricultural extension and peacebuilding. Services provided by the public sector can be particularly difficult to measure, even though they may have substantial long-term benefits. Furthermore, peacebuilding and extension both compete with other public services, requiring that value be attached to each. As a further complication to the measurement of value, as Unruh pointed out, circumstances and needs may change rapidly—from survival to crisis management to recovery to stability—requiring a continually morphing set of services rather than adherence to a strict model.

Kevin Brownawell, Interagency Professional in Residence at USIP, observed that Bell's definition of extension—providing knowledge to farmers so they can make positive change—also can be usefully applied to the role of extension in peacebuilding. In the context of peacebuilding, providing knowledge is more feasible than solving problems. Similarly, referring individuals to other institutions is more viable than an individual attempting to serve in the role of an institution.

Finally, Dale Johnson, Principal Agent and Extension Specialist at the Western Maryland Research and Education Center, emphasized the importance of commitment, motivation, and adequate resources. Without motivation, an extension agent cannot be effective. And without the funds to travel to farmers or even to make telephone calls, agents cannot do their jobs. Money needs to be specifically available for these activities rather than being allocated entirely to salaries.

4

Capacity Building and Training

During the final session of the workshop, participants divided into three groups to discuss (1) capacity building and training for extension personnel, (2) organizational change and institution building, and (3) the technological infrastructure needed to support extension activities directed at both agriculture and peacebuilding. This chapter summarizes the discussions of the first topic, and Chapters 5 and 6 present the other two. The summaries in these three chapters should not be seen as conclusions of the workshop or of the subgroups. Rather, they report issues raised in discussion to provoke further thinking about and exploration of the connection between agricultural extension and peacebuilding.

Discussions of capacity building and training focused on skills, legitimacy, and processes.

SKILLS

Participants agreed that the essential skill that extension personnel need is technical knowledge of agriculture—they need to be good agriculturalists. In addition, participants identified a range of other skills and attributes that extension agents need to do their jobs well, including cross-culture communication, project management, and knowledge of the local community.

Not all extension agents would be expected to have all of these skills. But these qualifications could form the basis for a curriculum, and individuals could choose from that curriculum based on their background and the situation in which they will be working.

The subgroup members also discussed the skills needed to reduce conflict, such as mediating or facilitating between parties or, at a greater level of involvement, negotiating settlements or resolving conflict. They acknowledged that acting in such roles requires an astute awareness of the conflict situation and how extension services could fit into it, and that such engagement could augment an extension agent's agricultural mission or detract from it.

Peacebuilding can require not just additional skills but additional time, and if an extension agent does not have enough time for it, the activity will not be sustainable. Rather than being responsible for peacebuilding activities as part of their formal job responsibilities, extension personnel may need conceptual models that further peace in the course of their extension activities. They also may need the skills and knowledge to work cooperatively within customary institutions and processes for managing disputes at the village level.

The acquisition of skills that will enable agents to address problems in both agriculture and peacebuilding requires training, which, among other things, should enable agents to understand how their technical work helps resolve conflict. Subgroup members emphasized the importance of experiential training, so that extension personnel are applying useful skills even as they are learning them. Trainees also need opportunities to reflect on their experiences with others to build their skills.

The discussants made a distinction between skills required by local extension personnel and those required by donor organizations (e.g., central governments, international entities, NGOs) to make decisions about investment decisions (Table 4-1). The skills required by local extension personnel and managers in donor organizations often overlap but are sometimes distinct. For example, both local extension agents and donors need to be able to identify local partners, but extension agents need particular skills to interact with these partners effectively. The group agreed that distinction applies across all three areas related to capacity building: skills, legitimacy, and processes.

LEGITIMACY

To be effective, extension personnel need to build legitimacy by fostering high levels of trust and credibility in their local communities, subgroup

TABLE 4-1 The Capabilities Needed of Local Extension Agents and Donor Organizations

Skills and Attributes		Factors Contributing to Legitimacy		Process-Related Factors	
Extension Agents	Donors	Extension Agents	Donors	Extension Agents	Donors
Facilitation	Analytic skills	Commitment	Knowledge of local conflict management methodology	Involvement of stakeholders	Donor coordination
Mediation/Negotiation	Ability to teach/train using many media	Peacebuilding knowledge	Support for links to relevant information	Identification of champions (local engagement)	Support for knowledge building
Brokering		Ability to access relevant knowledge			Identification of champions
Project management	Mentoring	Nonpartisanship			Partnering
Leadership		Good agriculturalists	Awareness of what partisanship is		Patience
Communication/Cross-cultural communication					Awareness of impact on sustainability
Commitment					
Organizational skills					
Analytic skills					
Training-the-trainer skills					
Methodological skills					
Experience-based (field) skills					

members emphasized. And to maintain this trust and credibility, their advice needs to be objective, useful, and nonpartisan. There are numerous components to the establishment of an agent's legitimacy; participants cited technical knowledge, credible and trusted local partners, motivation, and vision.

To improve agricultural yields in a particular region, extension personnel need technical knowledge of what will work in that region. This in turn may require new research on crop varieties and practices for the region. Such research is more easily conducted in countries where a strong linkage exists between extension and research institutions, as is the case in the United States, but may be more difficult in countries where such linkages are weak or do not exist (see Figure 3-1). Universities are also a source of training for extension agents, and weak linkages with these institutions can impede that training. Supporting university faculty to train extension personnel, either at the university or in the field, can be a valuable role for NGOs, national governments, and industry.

Extension personnel also can gain legitimacy by working with local people who are trusted and credible. Identifying these individuals can be difficult, but it is a skill that extension agents need. In some cases, these individuals may already have made significant advances; in others, they may be respected members of a community who are not yet involved in extension activities. They also might be people with an especially useful store of information, such as visiting experts or university researchers.

The perception of an extension agent as a member of the government may enhance or detract from the agent's legitimacy. If the government is perceived negatively by a community, the agent may have a hard time engaging in peacebuilding activities. But such an association need not be a factor if the agent's connections to local areas are strong.

An important attribute in creating legitimacy is commitment or motivation. If an agent is motivated simply by a paycheck or by having a government job, that person's legitimacy will be suspect. But if an agent's motivation is to improve a community, whether through agricultural or peacebuilding activities, legitimacy is enhanced.

In some countries, extension agents do not have the trust and respect they do in the United States. They also may not have extensive agricultural knowledge—for example, if they were recruited locally simply so that they would be more accepted by the local population. Extension agents from local areas probably know the language, culture, and best people with whom to work, but they may also have a vested interest in outcomes, belong to an elite, take a job for the wrong reasons, or be distrusted by the populace.

A final characteristic mentioned by subgroup participants is the need for extension personnel to have a compelling vision of the future. Communities in conflict often seem to live day to day since survival is such an immediate priority. An extension agent can help by laying out a desirable future for the community. Discussants acknowledged, however, that in practice few extension agents have the skills or resources necessary to fully realize such a vision.

PROCESSES

There was general consensus that the agricultural extension agent's ability to understand when and how extension work and peacebuilding fit together—in short, to understand the whole process—is key.

Extension agents should be able and willing to assess what is required for both extension services and peacebuilding. Agricultural extension agents often do not spend enough effort analyzing needs and the steps to meet them, subgroup participants said. To make such assessments, agents should be aware of cultural practices; they can learn much from the local population, both about agriculture and about conflict. Indeed, at some point, the bulk or all of the responsibility for analysis and action can devolve to local communities and away from extension personnel.

Problem statements that explicitly identify what is needed can build consensus and provide objectives for extension personnel. Because problems change over time, these statements should change as well to reflect new circumstances and a better understanding of a problem.

Conflict situations can be extremely complex, and the information needed to assess a situation scarce, requiring special expertise and access to information to enable effective conflict analysis. Extension agents must understand not only the drivers of conflict, but also the consequences of their actions in terms of the conflict; for example, agricultural improvements may exacerbate conflict if their benefits are unevenly distributed. So it is important that agents be able to assess whether a particular action will result in good or harm.

Extension personnel also need to understand how agriculture fits into a larger picture—to consider not only peacebuilding but also health care, the legal and political system, income distribution, and so on. They will have to be able to work within existing mechanisms for conflict resolution and augment them if necessary and possible.

In addition, an understanding of the local culture is critical. For example, the residents of an area may not perceive their situation as a conflict, whereas others may, requiring sensitivity among those who would offer to analyze a conflict. It is also important to understand that there is a distinction between "postconflict" and "postviolence" situations: a region may no longer be subject to violence although conflict remains pervasive.

As mentioned earlier, linkages are essential for a project to be sustainable and can multiply the effects of individual extension agents, especially when financial support comes from international donors rather than taxation by the central government. The needs in a conflict situation can be enormous, so many people must be on board for sufficient resources to be available. Furthermore, unless the efforts of individual agents are scalable, outcomes are limited to what single extension agents can do in their local communities.

Finally, for extension services to be sustainable, it is essential both that agents remain current in agricultural knowledge and that senior agents train and mentor their subordinates. Regular training in skills and knowledge relevant to farmer needs allows agents to remain effective. Effective agents, however, tend to be hired away by other organizations, so new personnel must continually be trained and be prepared to step in. Extension agents should mentor younger agents, knowing that succession is only a matter of time.

5

Organizational Change and Institution Building

The second subgroup examined the institutions with which extension personnel work. What changes are required in how agents are organized, supported, and resourced for them to engage in peacebuilding activities? Participants discussed options for working with ministries of agriculture, the need for decentralization, and the challenges of ensuring program sustainability.

OPTIONS FOR WORKING WITH MINISTRIES OF AGRICULTURE

Extension systems typically operate in a ministry of agriculture, and the subgroup began its conversation by talking about changes required in ministries of agriculture to support peacebuilding as part of extension services.

One option was for a ministry to officially adopt peacebuilding as part of its mission. Perhaps, as suggested by Jon Unruh earlier in the workshop (see Chapter 2), the ministry of agriculture could facilitate the role of extension personnel as intermediaries between customary systems and statutory systems. Another possibility is that the ministry of agriculture could undertake conflict analysis, especially to the extent that conflicts are affected by agriculture. Or a ministry and its extension officers could become involved in the reintegration of regions previously held by rebel groups, as occurred in Colombia (see Chapter 3), perhaps by demonstrating the competence and credibility of state actors.

But extension services need not work entirely through agriculture ministries. For example, the department of transportation may be involved to ensure that products get to market, or the highest levels of government may need to be involved for change to happen.

Alternatively, because universities sometimes enjoy a credibility that governments do not, they might serve as anchors for extension activities. But research, extension, and education often fall under different ministries, so there might be institutional barriers to support and collaboration.

Participants pointed out that in many countries the capacity of the ministry of agriculture is severely limited. Many do not have extension services or have a very narrow technical focus rooted in the Green Revolution. Others have poor records of being able to recruit, train, and support such services, resulting in ineffective and unsustainable extension programs. NGOs can provide support for ministries, but often such efforts are not sustained once an NGO leaves. Furthermore, changes in a ministry can take considerable time, whereas conflicts typically generate immediate needs that must be addressed in the short term to avoid greater conflict.

States that are corrupt or predatory may wish to keep agricultural producers poor and dependent. In these cases, political changes are necessary at the state level to create sustainable interventions for development or peacebuilding.

The capacity of individual extension agents also is limited, participants pointed out. Giving them responsibility for peacebuilding may detract from their principal mission with objectives that are impossible for them to achieve. Agricultural extension agents first need to provide information about agriculture. Peacebuilding can come after that. But participants conceded that the legitimacy of the peacebuilding depends on the legitimacy of the agricultural advice (as described in Chapter 4). And, as was emphasized throughout the workshop, peacebuilding need not be explicit or even conscious. Extension can serve the purposes of peacebuilding, regardless of whether it specifically focuses on that end.

THE NEED FOR DECENTRALIZATION

Central ministries need to allocate resources and make policy decisions, but centralized planning tends to fail, in part because it generally is too directive and ignores local needs. The subgroup therefore turned to the possibility of a decentralized system with the capacity to support local grassroots extension activities that have a peacebuilding component.

For example, the core role of extension personnel is to distribute information. The information typically involves agricultural productivity, but it need not be limited to agriculture. In addition, extension agents can empower the people they serve to seek out information on their own or to make their needs known to others.

The great advantage of extension activities is their potential to serve farmers' needs in a bottom-up fashion, through both geographic distribution and the ability to address expressed needs, subgroup members said. A farmer may or may not be knowledgeable about ways to increase productivity, and an attentive extension agent can tailor advice accordingly. Ideally agents could also try to identify and prioritize assistance for the most pressing problems facing farmers, such as land tenure issues following conflict or water management disputes during a drought. It is similarly helpful for agents to respond simultaneously to both short- and long-term needs, so that early success paves the way for long-term improvements. For example, a producer association can address both immediate needs and the longer-term issues involved in sustainable improvements.

Participants cited some examples of successful decentralized approaches. The Agriculture Technology Management Agency in India was designed to be accountable to farmers and their needs. Another decentralized approach that has resulted in successes is the extension-supported business cooperative; in Armenia, such cooperatives have become among the most trusted actors in rural communities.

A bottom-up approach need not be antithetical to top-down directives, subgroup participants pointed out. The extension agent typically represents the state and needs support from the state to be effective. The development of institutional structures in the community, whether agricultural associations, schools, or health clinics, requires both grassroots and government support. Although the state may at times be a negative presence, it also can promote positive internal changes in a community. For example, universities, NGOs, or the private sector generally are not able to step in and resolve conflicts in the same way that government officials can, though nongovernmental entities can make governments more aware of conflicts.

Support for extension efforts may be more effective in some countries than others. Successful extension systems tend to be decentralized, increasingly pluralistic, participatory, market oriented, sustainably financed, and technology enabled. For systems with these characteristics, investments are likely to be more productive than for weaker systems. At the same time, as was emphasized throughout the workshop, one size does not fit all. One possibility is to identify a menu of evidence-based possibilities for what should work in different contexts. Hard data about approaches that work and do not work can guide modernization of extension services and improve the fit between services offered and needs expressed.

Support from NGOs could offer a bridge between the immediate postconflict period and longer-term sustainable development. However, NGOs work better as servants of government than as replacements for it, a subgroup participant noted. Following conflict, the government may be perceived as incompetent or untrustworthy. Government representatives need to convince the people that the government is a trustworthy and competent institution concerned with their needs. Thus extension systems should be organized to ensure that the government (and not NGO partners) receives the bulk of the credit due for any successes achieved.

ENSURING SUSTAINABILITY

Extension activities need to be both stabilizing and sustainable, subgroup participants said, and sustainability is often enhanced by support from multiple sources: public, private, governmental, or nongovernmental. Multiple sources of support also can enable decentralization. In the United States, for example, the extension service in each state or territory is operated by a land-grant institution, which, in addition to local, state, or territorial funding, receives some federal funding that can be used to support local extension agents who respond to local needs through community structures. In India, local governmental and administrative structures are vital to the successful implementation of local extension activities.

But generally government support in particular is necessary to ensure that services are sustainable, coordinated, and backstopped properly, through not only the training and "re-skilling" of agents but also proper monitoring, evaluation, and quality assurance for the service put in place.

Government support is also important because local areas are often resource starved. Local authorities can be empowered if given authority for resources, including taxation authority. When the central state controls all the resources, it can be difficult to have a bottom-up and decentralized system characterized by innovation and responsiveness to local needs. Outside funding organizations may provide support for a decentralized system, but coordinating multiple donors, and combining their efforts with state efforts, can be a challenge.

6

Technological Infrastructure

The subgroup on technological infrastructure focused on ICT as the technological area most likely to have an immediate impact on peacebuilding activities.

ICT is rapidly becoming more powerful and less expensive. For the price of a tank of gas, an extension agent or an agricultural producer can buy and use a technology that provides tremendous communications and information capabilities. Although even inexpensive cell phones are still too costly for some farmers, prices continue to drop while capabilities, infrastructure, and users increase. Some countries have essentially skipped developing wired networks for communications in favor of wireless systems. Moreover, companies and some countries (e.g., China) also are investing in technology in developing countries in recognition of their productive potential.

Investing in the newest technology simply for the sake of technology is a mistake, cautioned one participant, but new technologies nevertheless have a large and expanding potential to contribute to extension activities with both agricultural and peacebuilding goals.

TECHNOLOGICAL CAPABILITIES FOR EXTENSION AND PEACEBUILDING

What qualities are needed in technologies used in efforts to promote both extension and peace? They should be trustworthy, in that users should know that the information provided is valid. They should be inexpensive to acquire and use, though they need not be free, since free things often are not taken seriously. They should support long-term capacity to improve both agricultural productivity and social stability. And they should broaden access to information for all people in a community.

Once a technology and its associated infrastructure are available, the question becomes what problems to address. For example, what is the information needed to solve a particular problem, whether it involves land tenure, water rights, credit, or technical information? Participants cited several examples of how ICT is being used for both agricultural extension and peacebuilding. In many places, farmers call on cell phones for prices of agricultural commodities in different regions, thereby maximizing their income. They can also call a voicemail number and record a question; an expert then records an answer that is available and accessible to all farmers. Cell phones are also used to take pictures of documents and upload them in a secure location so that records will always be available. And blogs on agricultural subjects are an example of the many applications of social media to extension.

In these and other ways, ICT provides access to legal and other kinds of information that are useful for farmers. Technology can thus supplement or augment the advice of an extension agent, providing information that a producer can use to increase outputs.

POTENTIAL EXTENSIONS OF ICT

The subgroup discussed what participants alternately labeled "Gandhian innovation" or "frugal engineering," in which a community is encouraged and supported to determine how best to use ICT to solve its problems and meet its needs. As the Arab Spring demonstrated, technologies often are used in ways that were not envisioned when they were created. In such cases, the provision of bandwidth and low-cost technologies can lead to innovation that applies creative solutions to local problems.

The use of cell phones in particular has become prevalent and adapted for both agricultural and peacebuilding applications. For example, if a farmer engaged in a land dispute draws a map in the sand and an extension officer takes a picture of it with a cell phone, the picture becomes a piece of evidence that can contribute to settlement of the dispute. As another example, a displaced person can call friends or family members to check on the status of a home region. Members of opposing sides could even talk with each other on cell phones about differences or possible points of reconciliation.

It is important that a technology platform be neutral in its application. It should not force users into making certain decisions or otherwise be prescriptive. Because literacy cannot be assumed, quality-assured video presentations are a valuable feature. ICT also enables extension personnel to report back on things they see, such as violence or particular agricultural factors, thus creating a positive information feedback loop. And agents can use technology to exchange information with each other, enabling the rapid dissemination of best practices and innovations.

To operate as peacebuilders, extension agents must demonstrate inclusivity, locality, and neutrality in their use and support of technology. They can do this by making information available to all potential stakeholders, customizing information services to reflect local conditions, and remaining neutral to maintain the trust of local community members. The power of ICT is its potential to create change while meeting these criteria. For example, an extension agent who learns of an impending food crisis can take steps to institute a local coping strategy. Or, during time of conflict, an agent can serve as an archivist for records that may be destroyed in war.

Notwithstanding the variety of advantageous uses of ICT, there are some important factors to consider. Technologies need to be upgraded periodically because they change rapidly. People, however, often require more time, especially if they do not have much technical experience or background. A further complication is the reliability of access to technology, as some areas may lack consistent electricity service.

But overall, subgroup participants pointed to the potential of even simple technologies to make a difference in agricultural production and conflict reduction, especially in areas where the basic elements of a technology infrastructure, such as a power grid, are unreliable. For example, radio or simple computers using low-cost video can be both sustainable and scalable. In this way, even very simple and inexpensive ICT can enable a more equitable distribution of information in a postconflict situation.

INVOLVEMENT OF THE PRIVATE SECTOR

The private sector is inevitably involved in the provision of technologies for extension activities, and this involvement can take different forms. For example, a company may provide a technology, perhaps with support from a government or NGO, as a free public service that the private sector can use to sell additional services. Such cross-subsidization has been used in many contexts and is particularly powerful given rapidly increasing ICT capabilities.

Many technology companies have outreach programs to gain customers and demonstrate their ability to be good partners for governments and the public. Companies often partner with government to do such demonstrations, but may distance themselves in conflict situations to maintain a more neutral stance.

7

Final Observations

Development, agricultural or otherwise, is inherently about long-term political and economic improvement, and peacebuilding is about shorter-term stabilization. The two have strong commonalities. For example, building the community structures that enable peace, such as strong producer associations and schools and civic organizations, also supports development. In that sense, extension can support both development and peacebuilding by simultaneously building capacity and providing a means of managing conflict.

In their concluding remarks, the workshop co-chairs emphasized two major issues associated with efforts to combine agricultural and peacebuilding activities in an extension system: collaboration and sustainability. The workshop brought together people from quite different worlds, they noted, such as technology development, agriculture extension, and peacebuilding. The interests and concerns of these groups overlap, but they also have differing experiences and expertise. Integrating these separate worlds and moving toward next steps will require continuing the conversations started at the workshop.

In addition, extension efforts, whether they target agriculture or peacebuilding, serve both immediate needs and long-term goals. Many tasks are beyond the capacity of extension personnel, but by making small, cumulative changes over extended periods, agents can have a dramatic and positive effect on both agricultural productivity and factors that enable conflict management and peacebuilding. Because of their capacity to build social capital in rural communities, extension agents have real potential to improve the economic well-being and security of farmers.

Appendix A

Agenda

Workshop on Adapting Agricultural Extension to Peacebuilding
of the
National Academies and
United States Institute of Peace

May 1, 2012

US Institute of Peace
2301 Constitution Avenue NW
Washington, DC

The goal of this workshop is to identify what peacebuilding activities could be delivered as components of existing extension services and what organizational modifications and new capabilities would be required to do so effectively.

The day divides into two halves. In the morning, we will investigate how resource conflict manifests in rural communities and how extension and advisory services have been used to affect such conflict. In the afternoon, in breakout sessions, participants will brainstorm what capabilities are required to support a peacebuilding role for extension.

The breakouts will address (1) what changes are required in the skills of individual extension officers, (2) what changes are required in the organization of extension systems, and (3) what technological innovations are required in order for extension officers to integrate peacebuilding into their extension activities.

8:00 a.m.	**Breakfast**
8:30 a.m.	**Welcome and Goals for the Day** *Ann Bartuska*, USDA; *Pamela Aall*, USIP
8:45 a.m.	**Conflict in Rural Settings** Conflict affects agricultural communities in multiple ways. Disagreement between communities on rights to land and water access can act as flashpoints to initiate conflict. Likewise, in the aftermath of conflict, returnees whether refugees or demobilized soldiers can create conflict by stressing a community's economic and social resources. What can be natural additions to an extension officer's activities to manage these destabilizing phenomena? Speakers: *Jon Unruh*, McGill University *Caroline Hartzell*, Gettysburg College *Michael Jacobs*, PEACE Moderator: *Pamela Aall*, USIP
10:30 a.m.	**Break**
10:45 a.m.	**Extension in Fragile Societies** Extension agents working in rural communities are typically representatives of government with the responsibility to provide information and guidance to improve agricultural productivity. What are the effects of conflict on agents' capacity to deliver such services and what examples exist of extension agents using delivery of such services as a means to manage conflict effectively? Speakers: *Mark Bell*, UC Davis *Jim Conley*, Civilian Response Corps *David Nisbet*, Agricultural Research Service Moderator: *Ann Bartuska*, USDA
12:30 p.m.	**Lunch**

1:15 p.m.	**Breakout Sessions**

Capacity Building and Training (Room B214)
Extension officers support farmers by communicating information, by providing access to resources, and by organizing farmers to get to market. Given the need to continue to support the technical dimensions of farming, what are the skills required for extension officers to manage conflict effectively in their communities?

Organizational Change and Institution Building (Room B215)
Extension systems typically operate within a Ministry of Agriculture. Following conflict, resources can be scarce, and coherent guidance even scarcer. To enable effective frontline peacebuilding activities by extension agents, what changes are required in how agents are organized, supported, and resourced?

Technological Infrastructure (Room B241)
Extension systems have used various communications technologies (rural radio, for example) to communicate with dispersed agricultural communities. Cell phone technology is transforming what can be communicated and has created the potential for two-way conversations. How should this and other recent ICT innovations be applied in extension to manage conflict?

3:30 p.m.	**Reconvene** *Ann Bartuska*, USDA; *Pamela Aall*, USIP
4:30 p.m.	**Adjourn**

Appendix B

Attendees

Workshop Co-Chairs

Pamela Aall
Senior Vice President
U.S. Institute of Peace

Ann Bartuska
Deputy Under Secretary
U.S. Department of Agriculture

Steering Committee Members

Cathy Campbell
President and CEO
CRDF Global

Mark Epstein
Senior Vice President of Development
Qualcomm

Brian Greenberg
Director of Sustainable Development
InterAction

Mike McGirr
National Program Leader, NIFA-USDA
U.S. Department of Agriculture

Donald Nkrumah
Associate Director, Global R&D
Pfizer Animal Health

Riikka Rajalahti
Senior Agricultural Specialist
The World Bank

Expert Participants

Gary Alex
Farmer-to-Farmer Program Manager
U.S. Agency for International Development

Tom Bamat
Senior Technical Adviser, Justice & Peacebuilding
Catholic Relief Services

Lawrence Barbieri
Special Projects Officer, Reconstruction & Stabilization/Military
U.S. Department of Agriculture

Mark Bell
Director, International Learning Center
UC Davis

Linda Bishai
Senior Program Officer
U.S. Institute of Peace

Kevin Brownawell
Interagency Professional in Residence
U.S. Institute of Peace

Jim Conley
Senior Agriculture Adviser, Civilian Response Corps
U.S. Department of Agriculture

Mike Deal
Executive Director & CEO
Volunteers for Economic Growth Alliance

Montague Demment
Associate VP, International Development
The Association of Public and Land-grant Universities

Ruha Devanesan
Executive Director
Internet Bar Association

Rikin Gandhi
Chief Executive Officer
Digital Green

Caroline Hartzell
Professor of Political Science
Gettysburg College

Sheldon Himelfarb
Director, Center of Innovation: Science, Technology and Peacebuilding
U.S. Institute of Peace

Julie Howard
Deputy Coordinator for Development, Feed the Future
U.S. Agency for International Development

Deborah Isser
Senior Counsel, Justice Reform Practice
The World Bank

APPENDIX B

Michael Jacobs
Chief of Party
Afghanistan Pastoral Engagement, Adaptation and Capacity Enhancement (PEACE)

Dale Johnson
Principal Agent and Extension Specialist
Western Maryland Research and Education Center

Susan Johnson
Associate Director,
Afghanistan Pastoral Engagement, Adaptation and Capacity Enhancement (PEACE)

Walter Knausenberger
Senior Regional Environmental Policy Adviser
U.S. Agency for International Development

Cindi Warren Mentz
Director of External Relations, Middle East and North Africa
CRDF Global

David Nisbet
Supervisory Microbiologist, Agricultural Research Service
U.S. Department of Agriculture

Judith Payne
e-Business Adviser
U.S. Agency for International Development

Siddhartha Raja
Analyst
World Bank

Brian Rudert
Chief of Party, Afghanistan Capacity Building and Change Management Program
Volunteers for Economic Growth Alliance

Frederick S. Tipson
Special Advisor, National Academies–USIP Roundtable on Technology, Science, and Peacebuilding
U.S. Institute of Peace

Jon Unruh
Associate Professor
McGill University

Jerry Upton
Consultant
Qualcomm

Maureen Whalen
Senior Analyst, Office of Science & Technology
U.S. Agency for International Development

Jacqueline Wilson
Senior Program Officer
U.S. Institute of Peace